BEARS of the North

Steven Kazlowski

[handwritten inscription:]

Kaztowski Rocks

To Eliška
you have
a great
Dad
be well

hancock
house

ISBN 0-88839-591-4
Copyright © 2005 Steven Kazlowski

Printed in South Korea—Pacom
Production: Rick Groenheyde, Laura Michaels
Editing: Ingrid Luters

Cataloging in Publication Data

Kazlowski, Steven, 1969–
    Bears of the north/ Steven Kazlowski.

    ISBN 0-88839-591-4

    1. Bears—Canada. 2. Bears—Alaska. I. Title.
QL737.C27K39 2005   599.78'0971   C2005-900146-1

Published simultaneously in Canada and the United States by

HANCOCK HOUSE PUBLISHERS LTD.
19313 Zero Avenue, Surrey, B.C. Canada V3S 9R9
(604) 538-1114 Fax (604) 538-2262

HANCOCK HOUSE PUBLISHERS
1431 Harrison Avenue, Blaine, WA U.S.A. 98230-5005
(604) 538-1114 Fax (604) 538-2262

Website: www.hancockhouse.com
Email: sales@hancockhouse.com

Contents

Dedication

This book is dedicated to Merlin Traynor and Walt Audi for their kindness and hospitality that is unmatched by any hug a bear could ever give.

Foreword

Some years ago I was at the Alaska Airlines terminal in Deadhorse, Alaska before checking into a flight to Anchorage. A rugged and worn young man I noticed piqued my curiosity as he stood outside next to his station wagon piled high with gas tanks. A conversation was struck up. He was a young wildlife photographer flying by the seat of his pants. His name . . . Steven Kazlowski.

Many seasons have passed since that initial meeting, and since then Steve has discovered the hospitality of the community of the Arctic. The oilfield workers in Prudhoe have adopted him. Bush pilots and Inupiat give him shelter, work and collective wisdom. The community of people, the people who live in the small native villages and the workers that come and go from the oilfield, give of themselves. Nothing is really expected in return for the hospitality. Perhaps they all wish that the visions and experiences they have had in this spiritual landscape will be expressed in Steven's reverant photography.

And then there are the bears.

Bears are creatures that are like us in many ways. They have a culture that defines how and when they act—frequently in explosive outbursts of energy. Bears have different personalities. They display individual quirks and eccentricities in the tender moments between a mother and her cubs and in the violence of huge boars as they clash over food or a potential mate. Even 1,500-pound male polar bears will play tag on the ice. No, they are not so different from us after all.

Steven Kazlowski has watched the polar bears play tag like children, recognizing each other from year to year. He has observed the power of a polar bear pulling a seal, weighing over a hundred pounds, through a breathing hole with only its mouth. He has seen great brown bears clash with amazing force only to lumber away from each other.

Steven has observed and photographed wildlife from southeast Alaska, through the interior of Denali National Park to the stark and serene Canadian Arctic. His travels have put him on a small sailing boat across the Gulf of Alaska fighting for survival in a huge storm. He has survived by his wits and his passion for the light that illuminates the lens of his camera. He has spent countless hours, days, weeks and months waiting and watching. The hardship of his daily life often included one or all of being cold, tired, wet and hungry.

Wildlife in the north is difficult to observe and photograph. Distances are huge. The perfect light is elusive. This book is a photographic log of Steven's journey. Almost all of the pictures were taken without the use of a telephoto lens. Steven's gift to us is a world that most of us will never see. I stand in awe of the beauty and immediacy of his photos.

— Bill Morris

Black Bears

Black bears are an amazingly variable and adaptable species. They have been able to hold onto much of their former range and extend themselves into what was once the grizzly bear's (brown bear) range. Out of the three bear species in North America, the black bear is the only one that is at home in the trees.

A black bear sits back and takes a break from eating spring vegetation.

A young female black bear climbs high in a tree to seek safety as other bears pass by on the trail below her. Black bears are typically shy and easily frightened.

This jet-black bear pauses as it makes its way down to Anan Creek to fish for pink salmon. Western black bears range in color from jet-black to brown, cinnamon and white. Even silvery-blue colored black bears are known to exist in different places throughout North America.

A rare white Kermode bear, also known as a Spirit bear, and her black cub wander along a tidal shore of British Columbia looking for a salmon meal.

A spring black
bear cub sits
high in a tree
crying out for
its mother.
It quickly
climbed the
tree to
hide from
approaching
bears.

A young female black bear rests on an old growth log in the rainforest. Black bears are quite at home among the trees of the rainforest whether it be a standing or fallen tree.

A black bear sow and her two spring cubs roam the hillsides of Exit Glacier grazing on new vegetation in Kenai Fjords National Park. Black bears usually give birth to two or three cubs every other year during the mother's hibernation.

A Spirit Bear sow demonstrates to her black cub how to catch chum salmon. The beautiful white Kermode bears are commonly known as spirit or ghost bears. They can be found from southeast Alaska to the central British Columbia coast.

A white black bear is called a Spirit Bear.

Along the central British Columbia coast and surrounding islands, the Spirit Bear is a unique subspecies of the North American black bear. Approximately one in every ten bears is white or cream-colored due to a recessive gene. Other Kermodes are all black.

A black bear sow with her cub in the rainforest of Anan Creek.

A black bear sow coaxes her young cub in a tree that it is safe to come down.

While the larger grizzly/brown bears are more comfortable fishing in open streams and rivers, the shyer black bears tend to look for nooks and crevices to grab fish. This black bear emerges from between boulders with a lucky catch.

A rare blue glacier bear is another color variation of a black bear.

This female Spirit Bear makes its morning feeding rounds at low tide looking for salmon.

Spirit Bears thrive in their rainforest home on a diet of green plants, berries and salmon. In the winter, they hibernate in dry cavities of giant old growth trees, protected from winter storms.

Along the steep banks of Anan Creek, bald eagles join black bears in a feast from one of the largest pink salmon runs in southeast Alaska.

A female Spirit Bear and her black cub
walk on a log along the high tide line.

pointed head helps perfect its skill of pulling its main prey, seals, out of their breathing holes. Adults can reach 1,500 pounds and twelve feet in height.

Under an Arctic sunrise, a polar bear cub gets breakfast from its mother. Polar bear cubs lack adequate layers of fat and their mother's milk gives them energy to keep warm.

Polar bears are adapted to life on sea ice. Their white coats have water repellent guard hairs, dense underfur, and fur nearly completely covering the bottom of their feet. Polar bears are as much at home in water as they are on ice or land. Polar bears are excellent swimmers and can stay underwater for a minute or two. They can dive to a depth of fifteen feet and swim up to six miles per hour. They travel great distances along drifting pack ice.

Young bears play constantly, sharpening their hunting and survival skills, and show their love for fun. This young bear coaxed its mother into a game of tag, in and out of the water, which lasted for six hours.

Polar bears' instincts allow them to know the same location or path on the pack ice that they travel each year.

Polar bear paws are very large; they have coarse sandpaper-like pads and are covered with fur, which keep them from slipping on the ice. The claws also make it easier to walk on slippery ice.

These polar bears nap as they patiently wait for the sea ice to freeze so they can begin hunting for seals, something that will become more difficult as global warming changes their Arctic environment.

Polar bears' innate love of playing sharpens their skills for survival in the Arctic. This bear tests its ability to break holes in the ice by pouncing on it, as it will do later in life to grab a seal out of its lair. Polar bears will sometimes pursue seals from under the ice. The ring seal is the staple of the polar bear's diet.

Polar bears appear to be highly social animals with an ability to maintain relationships with one another over a lifetime, but it does not mean that they will not be extremely aggressive towards each other when protecting their young or a limited food source.

This female polar bear stands to a height of twelve feet as she checks the surrounding area to make sure it is safe for her and her cubs.

Two polar bear sows and their cubs place frozen meat into the water through holes in the ice to soften it before eating.

Even though polar bears look white, their hair is really made of clear, hollow tubes filled with air. This hollow hair helps direct the sunlight to the bear's skin, acting like a solar heat collector. Polar bear skin is all black which also helps to hold the heat from the sunlight.

Polar bear sows generally allow their cubs to play with other bears' cubs, demonstrating their social nature. This visiting cub plays a little too roughly for this mother's liking and she chases the little bear away.

These two polar bear cubs are not siblings but play together as friends while a mother, on the right, keeps a watchful eye and yawns. Yawning can be an expression of stress.

Two young polar bear cubs play what appears to be a game of paddy cake with one another.

A young polar bear practices chin-ups on the tip of a whale jawbone.

After killing a seagull with a swift swipe of its massive paw and eating its prey, this young polar bear proceeds to gently play with the gull wing tossing it into the air. This allows the bear to practice balance and controlled, delicate use of its front paws, which can be deadly accurate.

A young polar bear wakes up from a nap and stretches.

Polar bears, like this one pulling its head out of a hole in the ice,
can drag 100-pound seals out of their breathing holes.

Polar bears are curious animals that, when watched, will often investigate what is watching them.

An area of slush that has frozen overnight creates a slick ice skating rink which this mother bear and cub thoroughly take advantage of by "skating" together across the ice on their backs.

This photo shows a good example of Alaskan polar bear territory along the Arctic coast in the Arctic National Wildlife Refuge with the Shublik Mountains in the background.

Two young cubs curiously sniff my scent from the air.

The stance of this polar bear is a clear warning of aggressive behavior—its head turned and down, hissing as it looks for a clear approach.

A polar bear belly-flops in the ice to cool off. Overheating can be extremely dangerous to polar bears.

On a sunny afternoon, along the Katmai coast of the Alaskan peninsula, a male brown bear takes a siesta.

Grizzly/Brown Bears

Brown bears, also known as grizzly bears, inhabited North America before any human presence on the continent. The range of this majestic bear has been reduced to remnant populations for most of the continent. The remaining strongholds of grizzly bear populations lie in northern wilderness areas. To be in the presence of a grizzly is to be struck with fear and awe.

Two adult male grizzly/brown bears find time to cool off and wrestle in the water.

These two brown bears approach one another shaking their heads back and forth—the call to engage in play. Play can continue for hours but can also quickly turn into fighting. The lighter bear is no different from the darker bear other than it is an extremely rare blond color variation.

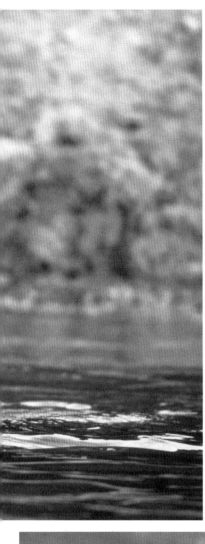

A brown bear
stands up in
a stream
looking for its
next salmon
to pounce on.

Brown bears will spend hours wading through streams and rivers as they hunt seasonal salmon runs.

Brown bears also inhabit southeast Alaskan and British Columbian rainforests such as this one feeding upon salmon along the banks of Anan Creek.

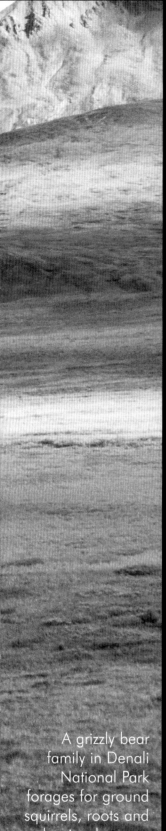

A brown bear sow and her cub watch for oncoming bears along a salmon stream in Katmai National Park.

Grizzly bear habitat can vary greatly: from rich coastal salmon streams to inland mountains, to coastal plains. This mother and her three spring cubs call the Arctic coast their home, which is the northern limit of the grizzly's habitat.

A grizzly bear family in Denali National Park forages for ground squirrels, roots and berries, because they do not have the large salmon runs to feed upon.

Salmon are a source of rich food in the coastal grizzly bear diet, which allows the grizzly to reach its large size.

Large male brown bears can eat over a dozen salmon in one sitting.

Young brown bear cubs, forced to keep up with Mom and overcome their fears, learn new strategies for survival.

A barren ground grizzly family walks the beach and lets their presence be known.

A barren ground grizzly bear wakes from its long hibernation in spring and is forced to look for carrion (dead animals) or to hunt mammals. In the Arctic the ground stays frozen and the bear cannot reach roots and spring vegetation.

Three young brown bears watch as their mother fishes.

Brown bears have different ways of catching salmon: some will dive for them, some will chase them out of the water and onto the banks and others, such as this one, will pin them down at the bottom of a streambed.

Two brown bears cross one another's paths as they fish for salmon, fighting for prime spots along the riverbed.

A young brown bear attempts to take a salmon from its mother's mouth in Katmai National Park.

On a lazy fall day along the Alaskan peninsula, this sub-adult female brown bear takes time out to stretch as it wakes from a nap, akin to bear yoga.

A brown bear sow and her three spring cubs cross a river at low tide in search of salmon.

It is a rarity to see grizzly bears and polar bears feeding in the same area as they are here. Little is known about how much the polar bear and grizzly bear habitats overlap along the Canadian and Alaskan Arctic.

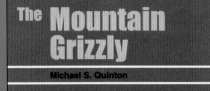

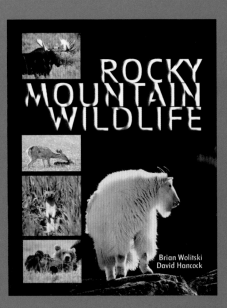